A LOOK INSIDE YOU

HUMAN BODY FOR KIDS

by Dr.A

AF719994

Human Body for Kids

Science for Kids, Volume 1

Dr. Amrita Basu

Published by Amrita Basu, 2025.

HUMAN BODY FOR KIDS

First edition. October 17, 2025.

ISBN: 979-8232665098

Written by Dr. Amrita Basu.

Table of Contents

This book is dedicated to my daughter.

HUMAN BODY FOR KIDS

A look inside you

by Dr. Amrita Basu

HUMAN BODY FOR KIDS

A look inside you

By

Dr. Amrita Basu MBBS,MS

Edited by Dr. Abhijit Misra MBBS MD

Published by Amrita Basu

First edition 2024

https://healthwealthbridge.com

Disclaimer

While all attempts have been made to make the book a guide to help address the subject, it is meant as a resource to help and inform. Since health and healthcare is a constantly evolving subject, the author, editor, or publisher individually or together will not be held responsible for any errors of omission and/or commission. This information is meant to promote health but is not to be taken professional medical advice. Kindly consult your child's doctor before making any changes to his or her diet. This book is not a substitute for a doctor's consultation.

Introduction

I wrote the chapters of this book during the pandemic shutdown of 2020.I feel deeply about access to education and books online and offline are the best ways to augment the learning of curious kids.

When schools were shut, it was upsetting in more ways than one. But that's not the topic I want to talk about today. I want to talk about the trouble I had getting into an age-appropriate human biology book for my then nine-year-old. Talking age-appropriate is crucial. You don't want to gross them out or give information overload.

"Help will always be given at Hogwarts to those who ask for it."

— J.K. Rowling, Harry Potter and the Deathly Hallows

About the Author & Editor

Dr. Amrita Basu is an ENT surgeon, previously a teacher in a medical college, and a mommy to a little girl. She started in 2014, when she realized she had a passion for sharing healthy lifestyle stories. Though she will forever be a Kolkata girl at heart, she shares a special relationship with Delhi and now lives in a small town in India, with her husband and daughter.

If she is not researching her latest book,or writing you can find her reading, pottering about her balcony garden or playing with her daughter.

Dr. Abhijit Misra is a paediatrician and previously a teacher in a medical college. He is a dedicated doctor and loves supporting his partner in real life in all her "help the world" schemes. If not for his help, it would be difficult for Dr. Amrita to manage home, writing and work with equal passion.

They are passionate about spreading the message of health, wellness and healthy lifestyle. They love helping people design a healthy, happy life.

Anatomy of a Human: What's inside?

Maths, languages, physics, chemistry, history, geography are essential.But knowing what's happening inside your own body is crucial for keeping it healthy and happy .The body you are born with is going to be there with you for the rest of your life ,so take good care of it.. Most of us are born with all body parts working perfectly fine. Health and wellness are prepacked, luckily for the majority.

But what we do affects our body, and that's why we need to know what's what and how your action can alter your health.

Anatomy is the science of the physical structure of the body. Many systems make up our body, the bones, muscles, organs. These work together in tango and perfect sync to keep you functioning. The functioning part is called physiology.

It's not a simple job, and even if you find all the elements making up a human body, it's difficult to make an actual live human from scratch. We still need a stem cell to create bits of tissues. But to create the tissue from carbon, nitrogen, oxygen, hydrogen, Calcium basic elemental blocks, are not yet possible. Modern science can't still create a basic human living cell from scratch!

You get the picture right?

Fun fact: The human body of an adult man contains approximately 15 trillion cells.

When we talk about anatomy, it's a talk about what makes up the nitty-gritty of the body in a big picture sense. Human anatomy is a serious business and is a topic for first-year students in Medical College. But having a good idea about what makes your bodywork is crucial for taking better care of it. You may or may not want to take care of other people's health, but taking care of your own health is a must for a happy, healthy life.

So what makes up the Human Anatomy[1]?

Lots of systems, organs, nerves, veins, arteries, cells, and excellent organization.

The Human body is so well organized that you and I could take lessons in the organization from our own body.

Some parts you can see, like your eyes, nose, hands.

Starting from the brain, which is the control center of our body, each part is crucial.

Others you can't like your thyroid, your larynx, heart, brain, and the inside stuff.

Outside or inside, all of this is important. While a detailed study of human anatomy details is best left to medical students, I will take you on a journey inside the human body. It is a kid's guide, but I hope you will enjoy reading it as much as I enjoy writing it.

Online resources have scanty and sketchy information. It also doesn't give accurate information. That's why I am writing about the human body and what's happening inside. I hope this will help you talk to your kids armed with the right set of data. I will also share unique tips to keep it working in perfect condition.

---

1. https://healthwealthbridge.com/human-body-for-kids-video/

Brain and your Mental Health

What you expose your brain to is very important. The brain is the mastermind behind the working of all our systems. If your brain didn't work right, it would cause a lot of problems. Nothing works well when there are problems with the brain.100 Billion nerves and their connections(synapses) make up different functional areas of your brain.

Brain and your Mental Health

What is mental health?

It's your emotional, social and psychological wellbeing. Your brain needs to be anatomically, physiologically and biochemically normal, too.

Thought or Action? But what comes first? It may surprise you that both have an exciting way of working at the same time. The brain decides, and you do it at almost the same time.

Are you a right-brained person or a left-brain?

Ideally, it would help if you were using all parts. The right brain or left brain either may dominate, but you need a balance of both to have a happy, healthy life.

The anatomical , physiological, functional brain health and mental health all work together . The brain is the heaviest and most complex organ.

The aspect of brain health called mental health makes your brain you. The human brain is the reason behind personality, intelligence, behavior, body movement, and senses. If all the other part of your body is one side, your brain is on the other side in terms of complexity. Even then, it will not be a balance.The brain has a group of experts. Each important working independently and also interacting with each other in layered manners.

The main parts of the brain are:

Forebrain -cerebrum

Midbrain

Hind brain-upper parts of the spinal cord, cerebellum.

The cerebrum or forebrain is divided into lobes

Frontal lobes problem solving, judgment, and motor function.

Parietal lobes sensation, handwriting, and body position.

Temporal lobes memory and hearing.

Occipital lobes brain's visual processing system.

Brain nerve cells are called neurons. There are ganglion cells which support and help provide nutrition. The cerebrospinal fluid is like a bath of nutrients that nourishes the brain and protects it against injury.

To exercise your brain, you need to be careful about the work you make it do. Something which is challenging for you, daily gardening, exercising, writing a story or poem, learning a new language, learning a new instrument, embroidery can help you rewire our brain. It's a tool against dementia.

If you stop to think, that's your brain working again. Your brain truly never stops. If it did, it would be catastrophic. What it does is rest—a little when you sleep. That's why eight hours of sleep[1] is much needed for a healthy future.

You will hear people saying they function perfectly well with less sleep. But that's just now; wait for the sleep deficit to catch up, then you will understand all about the ageing brain.The best way to fight poor brain health and dementia is avoiding inflammatory plaque-causing food and challenging your brain.

If it's easy, it's not a brain exercise. When you talk about wellbeing, it's not just physical and social wellbeing but also mental wellbeing. Things that keep you peaceful inside are the things that are good for our well-being. Both nature and nurture play a role.

---

1. https://healthwealthbridge.com/regular-sleep-better-health/

What are the common problems which prevent wellbeing?

- Stress [2]can be a problem for brain health, too. Stress is famous for increasing the amount of age-related plaques in the brain.
- Substance abuse.
- Smoking.
- Sedentary habits.
- Malnutrition.
- High fat sugary, salty food.
- Poor quality of social interaction.
- Poor environment.

Brain protective and cardioprotective food are similar.

Is mental illness a problem of brain health?

Mental illness [3]is related to neurological or biochemical dysfunction of the brain and needs timely diagnosis and long-term care. These are problems that should be treated only by expert professional psychiatrists. It's a complicated area and involves multiple factors.

Two exercises which are very good for your brain health and are easy to do are walking and gardening.

Is there any physical exercise that helps the brain?

In the book, Ikigai[4], the author's research found that people who lived the longest with most functionality and agility ate the right food and walked and gardened every day.

---

2. https://healthwealthbridge.com/chronic-stress-32-solutions/

3. https://www.psychologytoday.com/us/blog/sacramento-street-psychiatry/201512/are-psychiatric-disorders-brain-diseases

4. https://healthwealthbridge.com/ikigai-antifragile-book-summary/

They also took care to talk to their friends and neighbors. Positive social chit chat has a neuroprotective effect! Walking[5] helps your brain too.

Green plants, even in pictures, have a calming effect abd helps to lower stress. A daily half-hour of gardening could be an enormous boost for your health overall, but more so for mental health. The microbes around nature are also beneficial for our health.When my mom wants to know why I am learning Madhubani art tell her I am exercising my brain.

That's what you should look towards, something difficult which you will enjoy mastering.

The education system previously was all about STEM education. I support it, but the additional A in STEAM is much needed. That's why art in education[6] can be a big help to learning.

Teach your children to express whatever they learn through a quick sketch or even a poem. Sometimes it will create strange art or funny poems. Whatever it is, it's going to help them be creative. We need to raise innovative thinkers. That's why only telling them to do stuff will no longer work. It's all about walking the talk and leading by example.

I never knew I could do it. The best part, all art is beautiful. Sharing is fun but not essential. What's important is you give your brain excellent exercise.

Diet and epigenetics:

Mental diseases have solid genetic correlations, but this role of the diet with epigenetics has opened up new research areas. Several innovative studies are pointing to the exciting possibility that the effects of diet on mental health can be transmitted across generations.

While I can't force everyone to eat what I eat or take up a hobby, I can talk about my experience with drawing, painting, lettering throughout 2020. It was one of the best ways to fight Coron[7]a[8]

5. https://healthwealthbridge.com/book-series/wellness-series/
6. https://healthwealthbridge.com/art-effect-on-kids/
7. https://healthwealthbridge.com/mommying-covid-times-lessons/

fatigue[9].When the going gets tough, the tough get going.So stop overthinking and start doing. Keep your brain healthy. Now is the best time, with the perfect atmosphere, when the kids are fed, happy, educated, settled will come, but later. Your writing or art needs to escape before you forget how.

You can save your brain? It's never too late to learn. Just Do it.

Other Resources you will like-

Diet and Depression https://healthwealthbridge.com/diet-and-depression/

7 Ways to Make Kids fall in love with Creative Writing https://healthwealthbridge.com/kids-creative-writing

Mental health for moms to be[10] Mental Health for Moms-to-be (Guide) - Healthwealthbridge[11]

Power food for the brain https://www.ncbi.nlm.nih.gov/pmc/articles/PMC4445595/

Linking what we eat to our moods: Review of diet, dietary antioxidants and depression https://www.ncbi.nlm.nih.gov/pmc/articles/PMC6769512/

8. https://healthwealthbridge.com/new-coronavirus-outbreak-in-china/
9. https://healthwealthbridge.com/mommying-covid-times-lessons/
10. https://healthwealthbridge.com/mental-health-for-moms-to-be/
11. https://healthwealthbridge.com/mental-health-for-moms-to-be/

Circulatory system for kids: Going round in circles!

The circulatory system is by default also called the **cardiovascular system**. The five litres of blood in your body travels approximately 12000 miles per day.

That's a miracle. But in our body, there is other stuff circulated too.

Circulation means going round in circles, recycling, reusing. Your body has the most efficient circulation possible. The balance of fluids and minerals is maintained by reasonable control of blood pressure and epithelial integrity.

In case of these not working, blood circulation can be hampered, resulting in shock. In simple terms, when I talk about a circulatory system for kids, it comes to this.

The circulatory system comprises-

Blood circulation: Heart[1], blood vessels like veins and arteries.

Lymph circulation: Lymph nodes, lymphatic vessels.

Veins: Carry blood towards the Heart. The venous system carries the deoxygenated blood full of CO2(carbon dioxide) to the Heart.

Arteries: Carry blood away from the Heart. Your arteries carry oxygenated blood full of nutrients to send to different parts of your body.Without it, the cells there don't get the crucial oxygen needed to live and work.

A fun fact: Notable exception to how veins and arteries work

From the right ventricle, the pulmonary artery takes **deoxygenated** blood to the lungs to get oxygenated.

The pulmonary vein carries **oxygenated blood** from the lungs to the Heart (left atrium). Do you understand the anomaly?

---

1. https://healthwealthbridge.com/cardiac-arrest-world-heart-day/

Lymphatic and CSF circulation?

Lymph circulation helps to collect extracellular fluids. All of this is finally empty into the Heart, the central organ controlling what comes in, getting it oxygenated, and then getting them sent to the different body parts.

Why is circulation important?

Just like you want to recycle the wastewater from washing your vessels in your kitchen for watering plants, your body recycles blood through the lungs to remove carbon dioxide.

Circulation through the lungs helps in removing carbon dioxide from the venous blood. It then gets filled with oxygen, becoming oxygenated blood. This is where the amount of Haemoglobin (a protein) you have is essential. The **Haemoglobin** in your red blood cells carries oxygen in combination with iron. In case your Haemoglobin is low, or you have low iron, or any other condition which prevents your Haemoglobin from carrying oxygen, there's going to be a problem.

In CO(Carbon monoxide) poisoning, the carbon monoxide molecule gets attached to the site of attachment of oxygen and the cells can no longer get their oxygen. That is the reason for CO poisoning deaths. Oxygen is crucial to life, and plants help you get it. So don't forget to take care of all your green friends for more oxygen to breathe in.

Did you know snoring or OSA[2] is a condition when your brain gets less oxygen?

The brain cannot survive for more than 8 minutes without oxygen, and good function is lost after 3 minutes. If resuscitation is done after four to six minutes of a massive cardiac arrest,[3] then irreversible brain damage and coma are often the results.

For good circulation, you need-

2. https://healthwealthbridge.com/snoring-serious/
3. https://healthwealthbridge.com/cardiac-arrest-world-heart-day/

- Healthy Heart[4]: Diet, exercise, and mental health[5] all play a role
- Good muscle tone, so exercise and diet again.
- Drink water
- Take care of your gut[6] too.

4. https://healthwealthbridge.com/happiness-and-heart-health-worldheartday/
5. https://healthwealthbridge.com/brain-your-mental-health/
6. https://healthwealthbridge.com/gut-feeling/

Digestive system for kids: What does it say?

This is a supersystem that lacks your brain and heart's glamor but is crucial for well-being. No one enjoys talking about acidity, reflux, constipation[1], and bloating[2]. It's not fashionable or polite. But as a doctor, this comes up for discussion with astounding regularity. People can complain of problems with gas(that's the exact word),causing everything from family trouble to headaches. Most of the time, they are not far wrong. What you eat and how you poop is crucial for good health.

I have written about the gut-brain axis[3]. About probiotics [4]and prebiotic, I have a book on Probiotics[5]. It's a short book that **distills the information t**hat has helped my gut and the advice I give my patients.

I had a nervous bowel and had a habit of taking too many PPI when I was in Medical college studying. My gastritis solution was a medicine that ensured that I always had traveller's diarrhea if I was not careful.

But few people who take OTC[6] antacids know that indiscriminate use can cause trouble in paradise. Lowers acidity- yes. But that secondarily increases the risk of poor digestion. You need stomach acid for digestion. But hyperacidity creates many problems.

Overdosing on PPI lowers the population of good gut microbes, known as probiotics. Increases risk of infections involving your digestive tracts.

---

1. https://healthwealthbridge.com/constipation-taboos/
2. https://healthwealthbridge.com/one-habit-fight-gerd/
3. https://healthwealthbridge.com/gut-feeling/
4. https://healthwealthbridge.com/probiotic-life-a-beginners-guide-to-probiotic-food-podcast-bonus-book-download/
5. https://healthwealthbridge.com/book-series/nutrition-secrets/
6. https://healthwealthbridge.com/careful-about-otc-drugs/

So what comprises the Digestive tract?

- Mouth
- Oral cavity
- Oropharynx
- Pharynx /throat
- Oesophagus
- Stomach
- Duodenum
- Small intestine
- Large intestine
- Rectum and Anal canal

All these parts together to help you take care of the food you eat.

If your system has too much gas(yes that's the word), your organs stretch, and you feel pain. If your gut has terrible bacteria, a nasty virus, or a lousy toxin, it can make you rush to the loo and throw up. You lose water and electrolytes; you feel weak, your muscles cramps, your circulation[7] is compromised. That's why diarrhea needs ORS or oral rehydrating solution.

There's a saying your eyes tell us about our soul-Your oral cavity tells your doctor about your digestion and chewing addictions(if any).Your poop habits determine how happy you are (all factors being equal)-this is simplifying the issue, but you know what I mean, right?

Toilet Ek Prem Katha is a super hit, and there's a reason Swachha Bharat started with a toilet.

Souchalay for the urban poor and rural households is much needed. For those of us who grew up in regular city houses, we don't understand this. But my *Sasural* in the village had a toilet outside the house when I got married. Now 13 years later, finally, an attached bathroom part

7. https://healthwealthbridge.com/circulatory-system-for-kids/

of a small renovation project. If you have ever visited the real Indian countryside, you know how precious that is.

Keep your digestive system happy by taking care of a balanced diet rich in fruits and veggies.

Don't forget to drink water and exercise, too.

All these work together to make you a happy pooper.

Excretory System for Kids :What's out?

Doctors have an unglamorous job of explaining bodily functions and what you can't do. Today I talk about the Excretory system for kids (and their parents). This is part of my human body for kids series.

The circulatory[1] and excretory systems have a close-knit relationship. When circulatory volume decreases, urine output is lowered. In hot seasons, if you don't drink enough water, your body will produce a smaller volume of high-colored urine.

What is excretion?

Excretion is the physiological process by which metabolic wastes are removed from the body. Any build-up of this means your body becomes a wasteland.

Excretion differs from pooping. The chemicals in your body filtered by your kidney as urine happens through the Excretory system.

What are the Excretory substances?

Excretory substances are urea ,uric acid, extra water ,carbon dioxide, salts.

What is the Excretory system made up of?

Its made up of Kidneys (a pair)
Ureter (a pair)
Bladder

What are the major disorders of the excretory system?

- Kidney stones [2]and ureteric stones
- Uraemic coma

1. https://healthwealthbridge.com/circulatory-system-for-kids/
2. https://healthwealthbridge.com/what-causes-kidney-stones-prevention/

- Cysts of the kidney
- Tumors of the kidney or bladder
- Nephrotic syndrome
- Acute pyelonephritis
- Glomerulonephritis
- UTI \Urinary tract infection

Now any problem anywhere your urine output is lowered ,like in Chronic kidney disease[3]. When you stop Excretion of these toxins your body is poisoned inside out it's called Uraemia. An uraemic coma is a serious condition.

If you want this silent Powerhouse to work, you must take good care. When it's working well, the only thing you need to do is drink enough water and go to the loo on time. Your pee hygiene is crucial because the urinary infection can reach into your kidneys too. That's why Indian-style toilets are hygienically safer especially if you are using a public toilet. It's a relatively no-touch toilet.

Products like pee safe[4] can help and do this if you are traveling and need to use a Western WC. In train, while traveling using the Indian loo

.

What are the specialized cells of the kidney ?

Kidney has specialized cells called nephrons which have filtering set up called Glomerulus. Any problem involving this causes a condition called glomerulonephritis. I have written about kidney stones[5] and why you should drink more water[6]. But I have not written about the Ureter and the urinary bladder.

Ureter

3. https://healthwealthbridge.com/chronic-kidney-disease-india/
4. https://amzn.to/3cVHIN1
5. https://healthwealthbridge.com/what-causes-kidney-stones-prevention/
6. https://healthwealthbridge.com/13-cool-facts-water-no-one-tells/

The Ureter is a specialized organ with specialized uroepithelium. If there's a stone in the Ureter, it's severely painful. The pain classically comes in spasms.

Fast facts:

The repeated UTI (Urinary tract infection)causes the structure of the Ureter which in turn makes it even more prone to UTI.

Surgery of ureter needs expertise and is done by Urosurgeons. Kidney and ureteric stones have stormy histories if not treated on time.

The bladder is another much-ignored organ. When pregnant, women feel their bladder getting full faster. The weight of the uterus with the fetus makes pregnant women pee more frequently.

Urinary frequency and burning sensation are also signs of UTI.

In women, if normal delivery is not done at an institution during child birth,with proper care risk of vesico-vaginal fistula, prolapse and other serious condition occur. These can cause incontinence and life long poor quality if life.

Is pooping also Excretion?

No, it's not. It's Egestion.

Fast facts:

The urinary system has in addition, the urethra, the genitals (different according to gender) is a part of the Genitourinary system.Other systems which are considered as having Excretory functions are skin,lung,liver intestine.

Different systems have specific specialized function.

Nocturnal enuresis -night wetting has a physiological aspect but will need treatment if not resolved. Continence or bladder control is achieved.

- 50 % 3 years
- 80% by 5 years
- 90% by 8 years

VUR-Vesico-ureteric reflux congenital condition.

Vesico-vaginal fistula or VVF occurs in women who have complicated (obstructed labour)home child births :the most common reason in developing countries.(Source)[7].It causes continuous leaking of urine from the bladder to the vagina. Medical and social problem causing very poor quality if life in the women affected.

---

7. https://www.karger.com/Article/Fulltext/499411#_853ae90f0351324bd73ea615e6487517__4c761f170e016836ff84498202b99827__853ae90f0351324bd73ea615e6487517_text_43ec3e5dee6e706af7766fffea512721_Results_0bcef9c45bd8a48eda1b26eb0c61c869_3A_0bcef9c45bd8a48eda1b26eb0c61c869_20Obstetric_0bcef9c45bd8a48eda1b26eb0c61c869_20cause_0bcef9c45bd8a48eda1b26eb0c61c869_2C_0bcef9c45bd8a48eda1b26eb0c61c869_20due_0bcef9c45bd8a48eda1b26eb0c61c869_20to_c0cb5f0fcf239ab3d9c1fcd31fff1efc_recurrent_0bcef9c45bd8a48eda1b26eb0c61c869_20fiistulae_0bcef9c45bd8a48eda1b26eb0c61c869_20in_0bcef9c45bd8a48eda1b26eb0c61c869_2031.1_0bcef9c45bd8a48eda1b26eb0c61c869_25_0bcef9c45bd8a48eda1b26eb0c61c869_20patients.

Genes and Genetic disorders: Human Body for kids

Genes: How important are they?

What's your eye color? What about your hair? And your height? Did you know it's possible to guess accurately how you will look and be by looking at your parents?

Have you ever noticed how similar you are with your siblings? That magic is the work of a magician called the **Human genome**.

All living things comprise cells. Each cell has a genetic material called DNA or deoxyribonucleic acid.

Computer code determines what program will run. Your genes determine how you will be, what you will do. Gene is the basic structural(physical) and functional unit of your hereditary plan.

Gene is a brief sequence of DNA, having multiple combinations of chemicals called bases.

Four different bases ATCG' Adenine, Thymine, Cytosine, and Guanine are found in DNA.

Some genes help code for specific proteins, but some don't. Every person gets two copies of each gene, one from each parent. That way, we are doubly sure that all the work gets done. The human genome project was ambitious and took years to complete. It mapped approximately 20,000 genes in the human genome.

Most of our cells, except reproductive cells (ovum and sperm) contain 46 chromosomes or 23 pairs. 22 pairs are autosomal. They are the same. The last pair is called the sex chromosome.

Watson and Crick are two scientists who discovered the double-helix model (like a spiraling staircase) of the human genome)

Most common karyotype compatible with life are :

- XX Female
- XY Male

- XO Turner Syndrome also called Gonadotropin dysgenesis
- XXY Klinefelter syndrome.
- XYY
- XXXY

The different gender identification topics are beyond the scope of this article and need another post.)

But the exciting part about genetics is its microscopic and crucial—both the sperm father and ova mother supply the genetic traits that a child will have. But the result is more than an addition product and is always **unique**. That's why siblings are different with certain common traits.

Most humans have 99% similar genetic traits. 98%of human gene match with the chimpanzee.

Both nature and nurture are essential for developing personality. Both physical and mental health parameters gave genetic components.

It's vital to get **prenuptial** screening done for thalassemia. If both parents carry the trait, then there's a 25% chance of the child having beta thalassemia major.

Genetic screening for diseases like Down's syndrome, Edward's syndrome are important during pregnancy. Addiction and substance abuse also have hereditary tendencies.

Colon cancer, too, has high genetic correlations, and that's why it's essential to know about genetic counseling. Preventive healthcare helps in such cases.

Genes for kids

Single gene disorder- Sickle cell anemia

Complex gene disorder: Certain cancers come in gene clusters, colon cancer

Chromosomal disorder-Changed, extra or missing parts of chromosomes like Downs syndrome .

Commonest genetic disorders in India[1] are

- Downs syndrome,
- Thalassemia,
- Sickle cell anemia,
- Cystic fibrosis,
- Tay Sachs disease.

A database for Indian genetic disorders 2.0 version is soon going to be available here http://www.igdd.iicb.res.in/

1. https://pubmed.ncbi.nlm.nih.gov/11262988/

Note from Author

Thank you so much for reading .When you turn the page ,Kindle will give you an opportunity to rate the book and share your thoughts on Facebook and Twitter .If you believe the book has helped you ,and worth sharing, would you take a few moments to let your friends know about it ?When it makes a difference in their lives, they will forever be grateful to you. I would also be eternally grateful if you posted a review on amazon . All the very best for everything you do .Keep shining

Dr. Amrita Basu and Dr.Abhijit Misra

More from Dr. Amrita Basu

Fruits for Life[1]

Picky Eater[2]

Probiotics & Life[3]

Nutrition Secrets & Life[4]

Childhood Obesity[5]

How to Walk, Write and Garden your way to a Healthy life[6]

7 Blogging Challenges to a Successful Blog[7]

Starting a Successful Blog[8]

---

1. https://www.amazon.in/gp/product/B0725SSB4G/
2. https://www.amazon.in/gp/product/B07FLS2Y1X/
3. https://www.amazon.in/gp/product/B07MGQW6LY/
4. https://www.amazon.in/gp/product/B07VSRDNY2/
5. https://www.amazon.com/product-reviews/B08KH2K65J/
6. https://www.amazon.in/gp/product/B0814FXX32/
7. https://www.amazon.in/gp/product/B07FLCH31Z/
8. https://www.amazon.in/gp/product/B07FM2PHQF/

Don't miss out!

Visit the website below and you can sign up to receive emails whenever Dr. Amrita Basu publishes a new book. There's no charge and no obligation.

https://books2read.com/r/B-A-HDHX-OLQGC

BOOKS 2 READ

Connecting independent readers to independent writers.

Also by Dr. Amrita Basu

Krishna Mishra Mystery
Krishna Mishra Mystery series collection 1(Books 1-4)

Nutrition Secrets
Childhood Obesity and Adult Life
Fruits for Life

Science for Kids
Human Body for Kids

Wellness Secrets series
How to Walk Write and Garden for the Healthy Life You Want

Watch for more at healthwealthbridge.com.

About the Author

Dr. Amrita Basu is an ENT surgeon, previously a teacher in a medical college, and mommy to a little girl. She started writing in 2014 when she realized she was passionate about sharing healthy lifestyle stories. Though she will forever be a Kolkata girl at heart, she shares a special relationship with Delhi and now lives in Malda town with her husband and daughter. If she is not researching her latest book, you can find her reading, pottering about her balcony garden, or playing with her daughter.

Read more at healthwealthbridge.com.

www.ingramcontent.com/pod-product-compliance
Lightning Source LLC
LaVergne TN
LVHW022325200926
848481LV00004B/1435